This book is dedicated to everyone at Little Tor Elementary School for always encouraging us to be the best version of ourselves.

Pearl Ridhi Dawadi

Once upon a time, there was a girl named Pearl whose dream was to become an astronaut.

One night, from her bedroom window, she saw a bright star shining into her room.

Pearl wondered if this was the wishing star that could make dreams come true.

After all, this was her only dream that would open
a magical way to the galaxy.

So she wished upon a star to fly and see what the space and planets would look like from up close.

"I wish upon the stars…"

"To see the magic on Mars…"

"To slide on Saturn's ring on a railcar…"

"To cool down on Venus with an ice cream bar..."

"To bounce on Jupiter's cloud like a rockstar…"

"To sing a song on Uranus with a guitar..."

"To run on Mercury like a NASCAR..."

"To come back to Earth with a dream to live not too far…"

That night, a fairy appeared and granted her the wish.

The girl, with protected armor, flew into space and was amazed by the planets.

She collected a rock from each planet and brought it back to Earth. It was a dream come true!

The girl eventually grew up and made her dream into a reality.

She worked hard and gave it her all.

She knew it was not going to be easy, but staying focused was the key she needed.

She was recognized and rewarded as the "most ambitious astronaut"!

Dream big, work hard, and believe in yourself. Know that you're enough to do what you set your mind to achieve. The day is not too far when you will shine like a star. Make your journey far, set your goals apart, and watch your dreams come to life.

Pearl Ridhi Dawadi

 # What is your dream?

_____

_____

_____

_____

_____

_____

_____

_____

www.ingramcontent.com/pod-product-compliance
Lightning Source LLC
Chambersburg PA
CBHW051835210526
45473CB00005B/1882